The initial toe bend being made with the essential long heat.

The completed toe bend. Everything flows from this and the shape should already conform with exact contour of the toe of the foot for which the shoe is intended. A good shoer must have the shape in the "mind's eye" right from the beginning.

The quarter is formed while still on the second heat and is "boxed off" with the hammer on the outer margin from quarter to heel.

erly shod to provide comfort in all sorts of working conditions.

This is the way it should continue to be today, but sadly it's not always the case. Perhaps I can be forgiven if I say one of the reasons is a lack of basic understanding of anatomy.

Marvel Of Nature

This said, it moves me to provide a brief study of the structure of one of the marvels of nature—the equine foot. To do this I am going to use as a model the Clydesdale horse, the native breed of Scotland.

This breed is the product of very selective mating during the eighteenth and nineteenth centuries by some very able horsemen. These men knew exactly the kind of animal they wished to produce and proved to be very successful. In every respect, the Clydesdale and its lower anatomy differs little or not at all from other native breeds.

As in every breed, the Clydesdale's foot can best be described as a "horny box." It contains various living structures which differ in form and texture. The foot can be divided into wall, sole, frog and white line. All of these are composed of horn fibres which are dead matter.

Horny Outer Wall

The horny outer wall is made of carbon, oxygen, hydrogen, nitrogen and sulfur and contains approximately 24% moisture. Growth stems from the coronary cushion and its functions are to carry weight and protect the sensitive parts within.

The horny sole is semilunar in shape and varies in thickness, being thickest at the toe and thinnest at the inside heel. The moisture content is about 37% and its functions are to take pressure in conjunction with the wall, as well as protecting the sensitive parts within.

The horny frog is somewhat similar in composition to the wall and sole, but somewhat softer in texture. It takes the form of an equilateral triangle and its functions are to prevent slipping, minimize concussion or jar to the limb because of its rubber-like qualities and help prevent contraction of the heels.

Along with the plantar cushion and lateral cartilages, it assists the circulation of blood through the foot. The moisture content is about 42%. In addition, the horny frog receives a waste-like substance from glands in the plantar cushion which makes it soft, yet tough and elastic.

White Line

The white line joins the sole and horny wall together, thereby cushioning a lot of concussion that would otherwise arise should the wall and sole be directly joined together. This is non-sensitive horn and it is here that special emphasis should be placed on the need for the driven nail that holds the shoe to start its journey up the outer wall.

The Periople

The periople is a sort of soft skin found on every healthy foot which comes about two-thirds of the way down on the outer wall. It gives the foot a smooth, shiny appearance, emerges from the periople ring and is secreted independently of the wall. It forms an extra protection to the moisture in the wall, thereby reducing evaporation.

Sensitive Foot

The sensitive foot is made up of wall, frog, coronary cushion, lateral cartilages, periosteum, ligaments, tendons, blood vessels, bones and nerves. They are all more or less sensitive and are enclosed in a sac called the corium which is a continuation of the demis or true skin. The part bearing the sensitive sole or frog is velvet-like in appearance which is due to the papillae that hang from it and from which growth is secreted to form new horn.

The surface covering the sensitive horn is drawn up into folds or leaves to form the sensitive laminae which dovetail with the horny laminae supplying growth to the horny laminae and the white line. The corium covers the sensitive foot in a similar fashion to a sock covering the human foot.

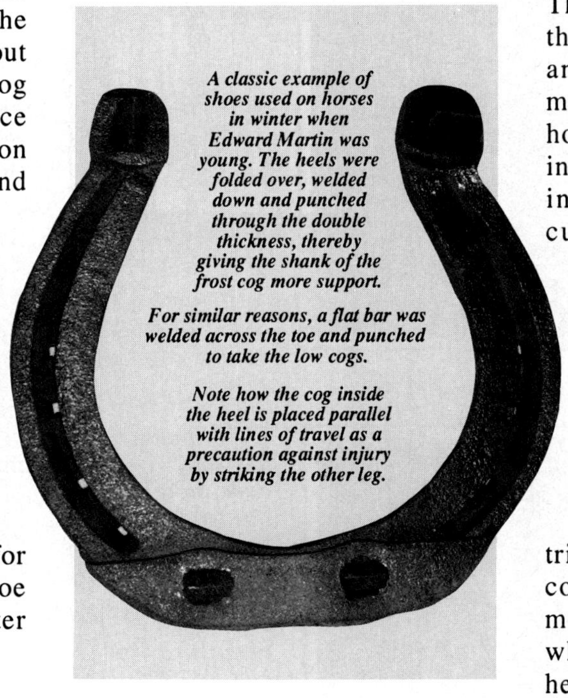

A classic example of shoes used on horses in winter when Edward Martin was young. The heels were folded over, welded down and punched through the double thickness, thereby giving the shank of the frost cog more support.

For similar reasons, a flat bar was welded across the toe and punched to take the low cogs.

Note how the cog inside the heel is placed parallel with lines of travel as a precaution against injury by striking the other leg.

Just as the human foot with its sock is covered with a boot, this is enclosed by the hoof. The corium is composed of dense connective tissue and is extremely vascular.

Plantar Cushion

The plantar cushion is composed of a meshwork of elastic tissues in which are lodged fat cells and fine elastic fibres. It is triangular shaped and lies between the two lateral cartilages, the end of the perforans tendon and the sensitive frog.

A vertical partition of fibres can be found running down the center of the plantar cushion from which other fibres string out sideways and become attached to the lateral cartilages. It is through this attachment that the back part of the foot is held together. The plantar cushion helps the horny frog carry out its function and also secretes a wax-like substance to the frog which makes the frog softer and rubber-like in texture.

Lateral Cartilages

The lateral cartilages found surmounting the wings of the third phalanx are two thin plates of gristle that extend upwards to protect the level of the hoof at the coronet. They form the basis upon which the hind part of the wall is molded and permit a certain amount of movement of the hind part of the hoof, being slightly elastic. It is intended that they receive pressure in conjunction with the plantar cushion—alternating as it does with the expansion/contraction of the foot in the natural movement of the horse.

As a result, the lateral cartilages help promote the flow of blood through the foot by acting as a kind of valve.

Every part of the body relies on the blood supply for its nutrition which is locked up in its corpuscles. These include proteins, minerals and other substances which are so necessary to maintain health and growth.

Anatomy Knowledge

Having dealt with the composition of the horse's foot and having described the functions of the various parts in a simplified, abbreviated form, I wish to make it clear I do not underestimate the knowledge of anatomy in general.

However, I have had cause to believe there are some farriers who have not taken the time or trouble to obtain this knowledge. I happen to think it is essential for every farrier to have a proper understanding of the art.

Practice without principles based on sound theory is like trying to pass an examination without being prepared. Knowledge of anatomy below the knees and hocks is essential because there are no superfluous parts; with each part working with its own function.

If one part ceases to function, extra work is thrown on the remaining parts and if all parts are to be equal, then they have to work in complete harmony.

Still on second heat, the outside branch is ready to fuller. The fuller is upright and "not lying on its back." Note length of fuller head, which helps to prevent "cremation" to the hand holding the fuller. A big hot shoe can reflect a lot of heat which causes considerable discomfort. An uncomfortable worker can lose the concentration necessary for good work. A good running fuller is invaluable and should be looked after like a brother. It should never need to be "pumped" to get the desired results.

Toe nail holes should be punched a little back from the start of fullering. This allows room to pull the errant nail without having to take the shoe off. The stamp should not mark the fullering only at the bottom.

Long heats are desirable, as hot as conditions allow, without defacing or burning the shoe. The heel is set at approximately 30 to 40 degrees, depending on the shape of the shoe. Note where the round face of the hammer is striking in relation to the edge of the anvil.

This knowledge plays a big part in the correct preparation of the foot for shoeing. It has to be combined with a knowledge and certain skills in the making and adaptation of the shoe to the foot to make the ideal union.

Back in the eighteenth century, someone wrote that "A proper mode of shoeing is certainly more important than the treatment of any diseases incident to horses. The foot is the part we are particularly required to preserve in health and if this art be judiciously employed, the foot will not be more liable to disease than any other region."

All of which only serves to confirm that there is little new in this world and the need for good care of the horse's foot and the attendant benefits of good shoeing practices have long been known and appreciated.

Color Counts, Too

In passing, it might be added there are a number of other factors, mainly hereditary, that can affect the soundness of the foot. With the Clydesdale one can correlate certain problems, both in the foot and lower limb, with color and pigmentation of the hide.

Light colored horses tend not to be so tough in the skin and have softer feet which are more prone to breaking.

In many cases, the light colored animal also is more subject to skin trouble when exposed to a lot of sun, mud or water. From my experience, bay in its various shades is the most serviceable color in a Clydesdale horse and the one least likely to give trouble with the feet or lower limb.

This is true with one provision—the texture of the hair. I have always found hard, curly hair on a Clydesdale's feather was synonymous with round soft bones and the inherent trouble it can bring when the horse is introduced

Big shoes demand a big fire which, in turn, demands big fire tongs. Edward Martin, shown here, says, "It is a fool that works in continuous discomfort caused by using short tongs only ideal for working at the anvil. A good worker should never be idle at the fire —always having a fire rake in his or her hands and always being in total command of their fire."

A big "clubby shoe" made many years ago. It was intended for a "street horse" and was a load in itself.

The quarter is now being formed. This part of the operation can also be carried out on the heel of the anvil. Note where the hammer is striking and the position of the tongs. At this stage, this will allow the forging of the bend to finish with the hammer striking solid almost on the center of the anvil horn.

Here is the branch forged out and shaped. Note that the heel (before bending) is in line with an imaginary line of travel.

The striker gives the heel the first bend. Note the angle of the hammer at impact. This leaves a slight indentation of the underside which locates the bend exactly where it is wanted. The same indentation also allows the worker to pull back on the tongs to get a close bend. Note how the tongs are in the line of pull and parallel with the face of the anvil.

The heel is being set down slightly. It does not take much if the correct amount of iron has been taken in. Note how the hammer face is introduced to the heel to provide an inward as well as downward motion. This provides a nice square finish to the heel.

The shoe is fullered almost in the center of the web (1 1/4 by 1/2-in.) and stamped in the second heat. It is nicely spaced with the outside heel nail hole being the width of itself beyond the half-way mark. Nail holes, with the exception of the heel nail which should be less than the others, are punched out at a pitch commensurate with the slope of the foot. This is assuming we are dealing with the average foot.

Note the negative pitch of the inside heel nail hole. Too much pitch in this area can spell disaster when nailing on the shoe.

to hard work.

This type of coarse hair also is more susceptible to infestation of a leg mite which causes the lower limb to itch intolerably. This can lead to many problems where the foot is difficult and dangerous to lift and almost impossible to hold when you do get it up.

It is therefore my firm conviction that fine silky hair is not only an adornment, but also complementary to other qualities.

Toe Vs. Heels

At the time the foal is born, the discerning owner is already looking for a nice shaped foot with plenty of size at the coronet along with good flat bone, etc.

However, the young Clydesdale can have congenital limb defects which have to be recognized and dealt with as in all other breeds.

Over the years, I have noted the average Clydesdale's foot has a greater predilection toward growth at the toe rather than high heels. This demands regular attention be given to cutting back the toe and rounding off the outer wall to get the animal back onto a proper sloping fetlock.

"Toe turned in" means the inside wall and heel must be reduced to maintain a level bearing surface.

"Toe turned out" demands the problem be dealt with in a similar manner, but in this instance the outside wall and heel both receive the same treatment. This is pretty straightforward.

Nowdays, the greatest proportion of Clydesdales are being bred for showing in Scotland and then a large number move on to the export market.

I personally do not know of a breeder who does not show. This sometimes provides a different train of thought from what used to prevail with the type of person who was only concerned about breeding their own replacements and keeping them right.

Show Ring Demands

The show ring means success and failure, often with selling

prices being sometimes proportionate to the awards gained. Unfortunately, this can lead to malpractice in shoeing in an effort to correct faults in the limbs and defects in the gait.

I freely accept it is not always enough just to provide corrective dressing of the foot and in some cases light plates have to be applied from a fairly tender age, such as 1 month.

There is nothing wrong with this in my mind as long as it is done sensibly and provides for a naturally balanced foot and leg.

What I am opposed to is the overemphasis on some aspects of gait that is demanded by some judges and the "stop at nothing" tactics adopted to attain this unnatural movement.

An example is the desire for excessive close action behind where the hocks almost rub one another. Nature never intended that this should take place, especially in draft horses with hindquarters and thighs commensurate with the vast build and conformation of the breed.

Even allowing for the fact that the open hocked horse will break down quickly when asked to move heavy loads, it does not mean to say the remedy lies in extremes in the other direction.

I have seen hind feet cut down on the inside branch until blood was drawn and the outside quarter developed through regular plating to a stage where it resembled a wing that had the same built-in resilience as an eggshell.

I have also seen three-quarter hind shoes applied to such feet where the outside branch was built up with extra iron to create an imbalance which had to be seen to be believed.

Square Toe Whim

Another whim is the demand for a square toe in front. I have been told this is done to provide an optical illusion so the foot, when advancing towards the beholder, gives the impression that it is big-

The inside branch forges out and the heel is ready for bending. The inside heel is slightly narrowered in the web and behind the heel. There should always be a distinction between the inside and outside branch of a shoe. The wall of the foot is at its thinnest at the inside heel and the shoe has to be fitted less bold than the outside. A broad web leaves a stone trap.

The fourth heat encompasses the shoe from quarter to quarter. This allows a quick assessment and for alterations to balance the rest of the shoe. The round hammer face should be run around the foot surface of the shoe from fullering to fullering. This will broaden the web at the toe to match the rest of the shoe broadened by fullering and gives the shoe a symmetry when viewed from the face side. Nothing looks worse than a shoe narrower at the toe.

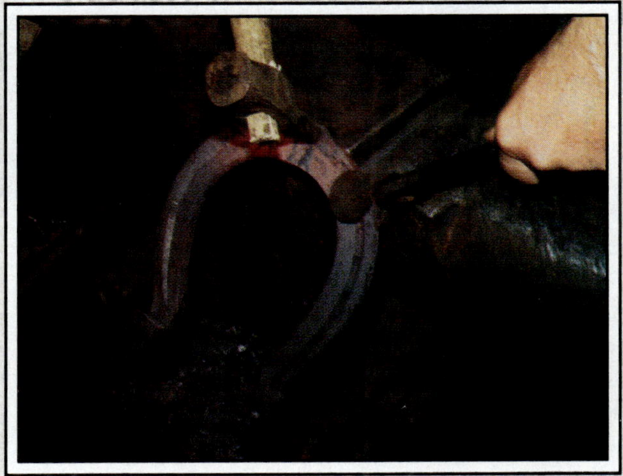

The clip should be drawn in the center of the fullering which should also represent the center of the shoe. The method shown here allows you to look down through the shoe and take in the whole balance of the shoe, including the heels. While a bob punch is used here, a clipper hammer could be used with equal success.

ger than it really is. Again, nature never intended that the shape of a draft horse's foot should become square across the toe.

Anyone who has watched a horse start off a load knows just how much the animal depends on the toe as it stubs it in to gain impetus.

Yet there are owners and breeders who would want this part of the hoof rasped back until the wall yields, almost with thumb pressure. It is a bad, bad practice and one which no self respecting farrier would want to be associated with.

Conformation Counts

The conformation of a horse is not only the product of the dam and sire, but also the environment. It can be altered by the farrier in the horse's first year of life.

Bone, as already described, is a solid object and cannot grow in overall length. As a result, all long bones have nature provide them with growth areas called epiphyseal cartilage which allow the bones to grow.

As the horse matures, the bone is subject to epiphyseal closure which involves the cartilage solidifying into bone. Consequently, whatever alignment the bones adopt at closure will remain as the confirmation of the horse for life.

It is when the epiphyses are growing—and only then—that they can be affected by pressure. If more weight is exerted on one side of the epiphyseal cartilage than the other, then that side will grow at a slower pace and thus alter the alignment of the bone.

As Robert Burns, Scotland's national poet, observed and wrote, "As the twig is bent, so also is the tree inclined."

It is at this stage that the farrier has an important part to play in getting the young Clydesdale to move straight and true. This is the time for judicious dressing and plating as no amount of the type of previously stated dramatic treatment done after the first year is going to alter that which is inherited at this juncture.

Any later attempt to alter the alignment of the bones can only alter the weight distribution down the limb and through the joints with subsequent discomfort.

By way of further illustration on the points I have made on "shoeing for show purposes," I have included some photographs along with a text explaining the various features.

At the same time, I am providing photos of typical traditional work shoes by way of complimenting my views on the everyday-type shoeing of the Clydesdale, whether on the farm or the street.

Starting Point

I will always favor a flat shoe, particularly in front if working conditions allow it. The problem in Scotland is we experience high rainfall where soft, slippery conditions prevail and where heels (called caulkins in Scotland) could be considered a necessary evil.

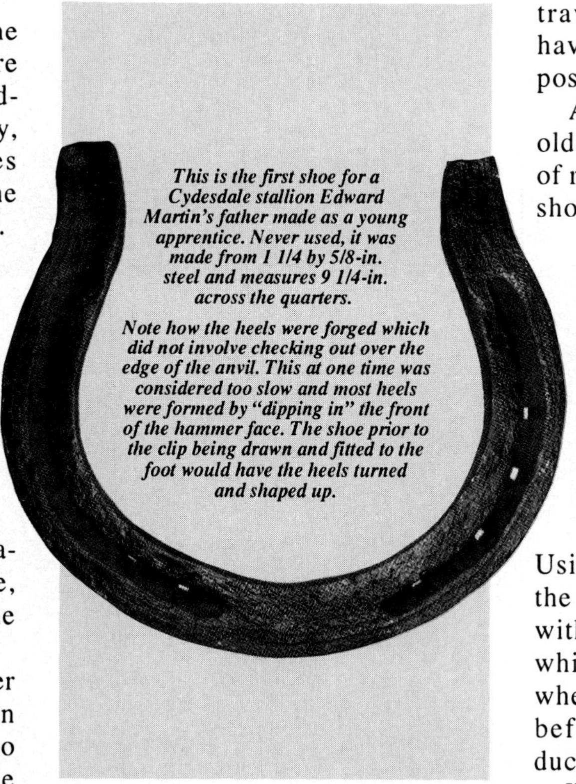

This is the first shoe for a Cydesdale stallion Edward Martin's father made as a young apprentice. Never used, it was made from 1 1/4 by 5/8-in. steel and measures 9 1/4-in. across the quarters.

Note how the heels were forged which did not involve checking out over the edge of the anvil. This at one time was considered too slow and most heels were formed by "dipping in" the front of the hammer face. The shoe prior to the clip being drawn and fitted to the foot would have the heels turned and shaped up.

Likewise on the street in the old days, the heel (or caulkin) was expected to penetrate between the granite slabs which at one time were used on lots of town and city streets.

By the same token, the toe piece on the hind shoe was intended to do a similar thing when starting off a load, for example. At one time these appendages on the shoe (toes and heels) were used as a means of competing with the hazards brought on by snow and ice.

This was known as "roughing" and involved taking off the shoes and sharpening both toes and heels before renailing the shoe. This was a time-consuming, laborious job, but one that had to be done if the horse was going to remain mobile in wintry conditions.

It was only a matter of time before frost cogs, which could be driven in and taken out at will, overtook this practice. But for many years, toe pieces and suitably sharpened heels proved quite successful in allowing the horse to travel when it otherwise would have been dangerous, if not impossible.

As a youth working with an older generation that could boast of real "iron workers," partly worn shoes were "bishoped." This was done when the horses came to the shop for shoe removal and refit. (Yes, this was done in those days by the thoughtful owner who considered it important not to let the feet become overgrown).

The shoes were put in the fire and heels were welded on individually and quickly. Using a simple but clever method, the toe received similar treatment without any detriment to the clip which was hammered up the sides when hot to thicken the material before welding heat was introduced.

With the toe piece welded on securely, the clip was redrawn back to normal. This fairly universal treatment meant the shoes contin-

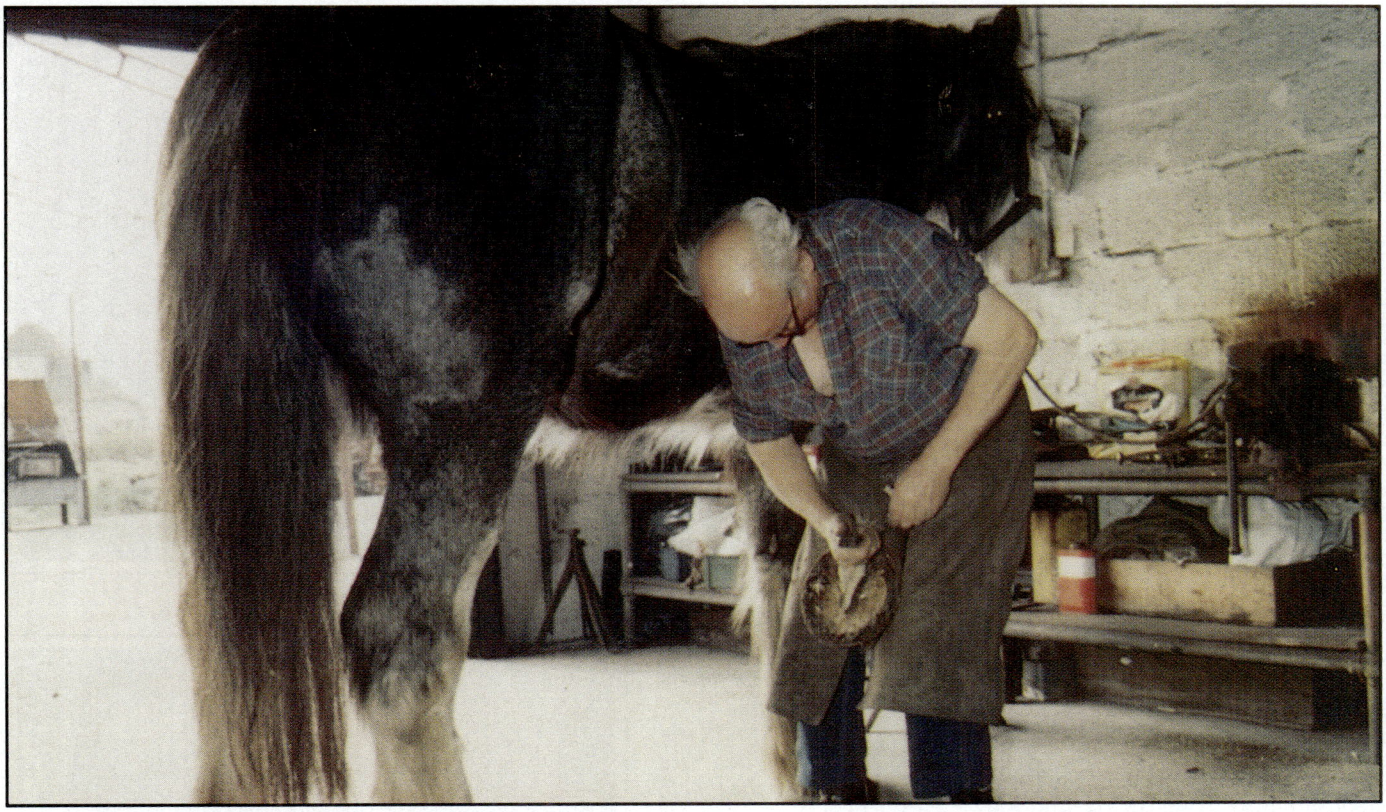

"My father believed an apprentice should make 72 pairs of shoes a day which was hard, hard work," says Edward Martin. "But it was amazing how you learned to keep a number of shoes all going at different stages in the fire."

ued to provide their function of enabling a good grip when needed.

To critics of this type of heels who say they are too "peg-like" and are subject to breaking, I have an answer. In 54 years of shoeing, I could easily count on one hand the number of breakages I have witnessed.

This brings me to the point of how they are formed or should be formed. First, the iron should be worked hot and bent hot with the correct set or angle to the web of the shoe achiéved with a minimum of hammering.

Heels (or caulkins) should be low and not started off as high and then hammered down. In essence, the skill lies largely in the experienced eye taking in the correct amount to form the heel.

Traditionally, Clydesdales shoes were almost always fullered, which allows for the extraction of the errant nail without sometimes having to take the shoe off. The type of nail normally used in conjunction with such fullering had a slightly slimmer shank which gave smaller holes in the wall of the hoof when driven.

It is also widely held that the fullered shoe has a distinctive appearance and the skilled workman obtains the visual satisfaction of a job well done. After all, beauty is in the eye of the beholder.

Clips, like heels, also must be sensibly done and commensurate with the shoe. They should be formed and drawn when really hot with not too sharp a point but still strong, especially at the root.

In everyday work, there have been times when I have seen fit to provide outside quarter clips as well as ones at the toe. If the horse is used in heavy work, there is no doubt in my mind that this additional clip helps prevent the shoe from moving and gives support to the nails.

Heavy Iron?

The use of too heavy a piece of iron in the making of the shoe should be avoided. It is counter-

Note module created by the bob punch has an indentation caused by the edge of the anvil. This will be taken out when the clip is finally drawn. Sharp edges on the anvil should be avoided when drawing a clip as it is important that it be strong, especially at the foot which should be slightly radiused on inside.

Note the position of these tongs which gives the worker a secure hold on the shoe and prevents any movement. Note angle of the hammer face which moves progressively towards the worker and finishes right at the point of the clip.

productive and can only cause the foot to hit the ground harder and leads to increased wear.

Where a 4-lb. shoe is used on a big horse in place of a 2-lb. shoe, it is a proven fact that the expenditure of muscular energy is sufficient on the four legs to raise 208 tons a foot high over and above the normal day's work.

In making a Clydesdale's shoe, I will always sacrifice a little thickness for width with its better cover. Having said that, I make sure the hammer is run around the inside edge of the foot surface of the shoe to prevent this part from impinging on the sole.

A good workman will always produce shoes that not only provide for every function of the foot but will also be "things of beauty" and a joy to behold. This shoe is balanced, clean, well worked and a product that gives pride and satisfaction as well as considerable comfort to the horse.

Preparing The Foot

Foot preparation on the Clydesdale for the reception of the shoe does not differ in any way from any other horse.

It is always necessary to see the horse led out to obtain a true picture of how the horse goes. An account has to be taken of how the foot breaks over, also its flight pattern, etc., before dressing and shoeing. Only in this way can balance and comfort best be achieved.

Assuming all is well with the break-over at the center of the foot

With clip at correct angle to the foot and before final leveling, back punching should be carried out. This should be done in the lightest possible form to shape the nail hole to exact dimension. Nail holes at this stage should have been pritcheled out at least three times from the face side. Outside nail holes are slightly wider apart than the inside holes.

After broadening web between fullering and after shaping the shoe, the clip should be drawn on the fourth heat. Again, the old fashioned clipping hammer is used which only requires one person. All that is required is to hit the same place repeatedly and not very hard.

"Work hard, work efficiently, work honestly and work humbly before God," says Edward Martin.

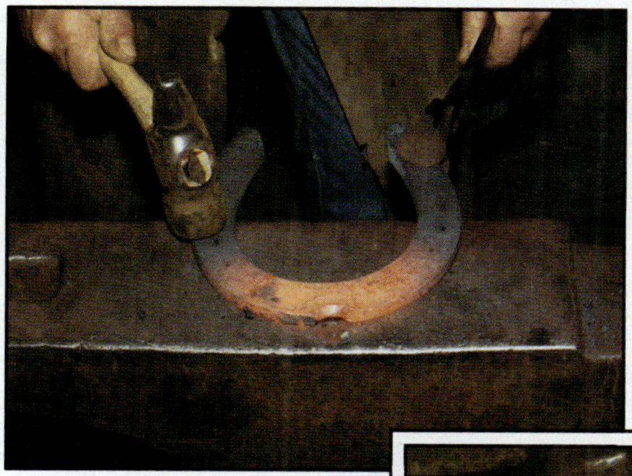

The nodule made with the clipping hammer is now ready to be drawn in a similar fashion to nodule created by the bob punch.

The finished article ready for fitting straight off the hammer and completed in four heats. Note the distinction between the inside web and the outside, the heel form and fullering. Spacing and pitch of the nail holes also differs considerably.

After back punching and before leveling, the round end of the hammer should be run around the inside edge of the foot surface to make sure it does not impinge on the sole when fit. This applies to most shoes with few exceptions.

and the flight pattern is straightforward, we can still encounter growth at the toe. This tends to give a wrong slope to the pastern and hoof when viewed from the side.

As stated at the beginning, this can be a tendency in the Clydesdale, and something that has to be watched. Viewed from the side, an imaginary line passing down the cannon bone to the fetlock should meet the ground behind the bulbs of the heels. The slope of the hoof and the pastern should be approximately 28 degrees and should not differ a lot from the slope of the shoulder.

To cope with the problem, I am a firm believer in letting the shoe back in the toe or alternately rasping the toe back to reasonable limits before fitting the shoe.

Flares have to be rasped off and development of straight horn encouraged. This has to be done at each shoeing because the horn tubules will have been influenced by the outward flare.

With each shoeing, the influence will lessen. Tubules that have been cut further up will not have been bearing any weight and will in due course straighten out. Because of this development, the flat foot deprived of its flares will start to regain its lost concavity which in turn will induce better heel growth.

Many years ago it was written, "That of all the faults of conformation of the feet, a weak heel was the worst because without depth and width of heel no horse will remain sound for long."

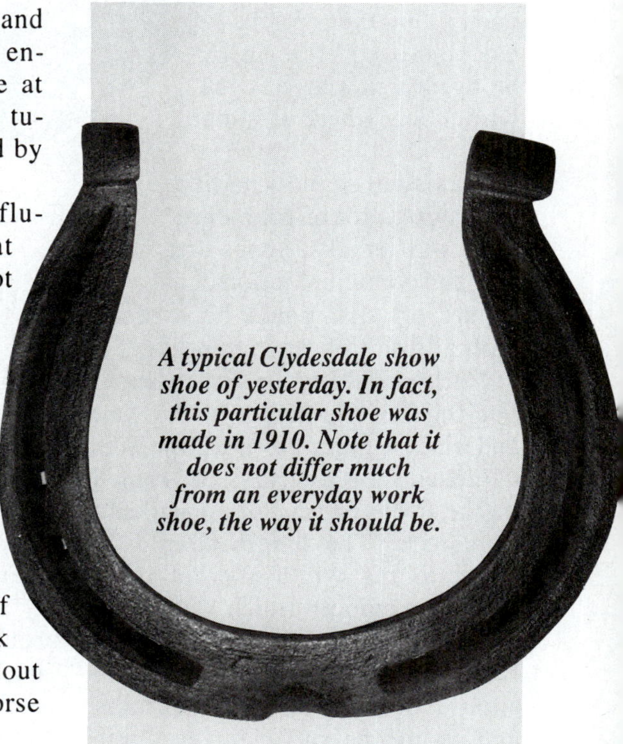

A typical Clydesdale show shoe of yesterday. In fact, this particular shoe was made in 1910. Note that it does not differ much from an everyday work shoe, the way it should be.

Nothing has changed and that statement still holds true today. Frogs should be trimmed, if necessary, of all ragged edges and cavities likely to catch debris. Attention should always be given to the seat of corn which is often forgotten about today until the horse actually goes lame.

In this connection, there are times when problems can be caused through the careless use of your pullers when taking off the old shoe and this should be watched.

I would stress regular shoeing at intervals running no longer than 6 to 8 weeks. By doing so, proper balance and a healthy foot can be maintained.

Watch Nail Holes

In conjunction with this method of preparing and fitting, much care should be taken in making shoes, especially with the nail holes coarse enough in their pitch. By so doing, the nail being driven starts at the white line where it should be.

Persistent shallow nailing can cause a lot of problems: hoof wall cracks, shoes are lost and a chain reaction can be set off that could be avoided with a little care and attention.

Nails should be driven to about one-third of the height of the wall and rise slightly towards the toe. Clinches should always be formed on the strongest part of the nail and be square. The length should be equal to the width and let in to provide a smooth finish.

It goes without saying that a large foot like the Clydesdale should be shod to provide room and should be sufficient in length to come back to the imaginary line, thereby providing the right kind of balance and support.

Life's Greatest Pleasure

In all my working life, there has been no pleasure to equal that of watching a well shod Clydesdale walk away from the shoeing shop door with a jaunty air born of a spirit of well being that almost seems to say thank you.

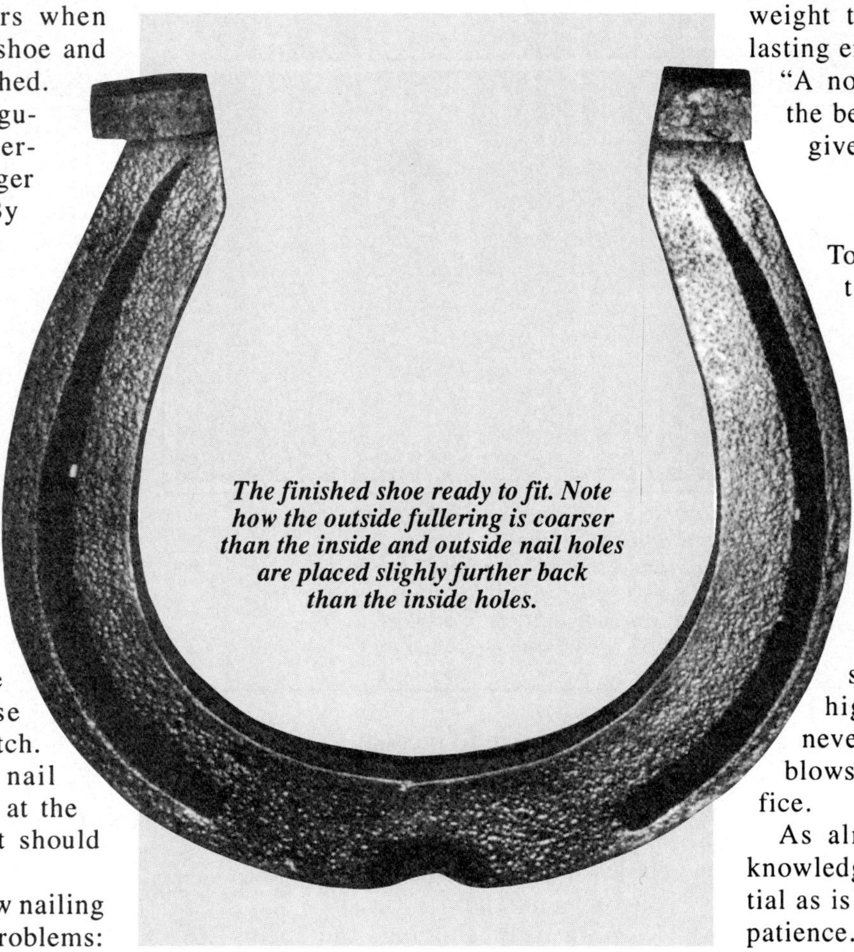

The finished shoe ready to fit. Note how the outside fullering is coarser than the inside and outside nail holes are placed slighly further back than the inside holes.

A Horse With Action

It has been written that "A Clydesdale is a horse with action, lifting his feet clean off the ground and not scuffing along. When going, the inside of every shoe must be made plain to the person standing behind.

"The foot must be open and round like a mason's mallet. The back should be short and his ribs well sprung. Quarters should be long and thighs well packed with muscle and sinew. He should have broad, clean, sharply developed hocks, broad knees in front with a nice sloping pastern set at an angle of 45 degrees from hoof head to fetlock joint.

"The impression created by a thoroughly well-built typical Clydesdale is that of strength and activity with a minimum of superfluous tissue. The idea is not grossness, but quality and weight that gives dignity and lasting endurance.

"A noble animal, deserving the best attention a man can give."

Summing Up

To give this deserved attention, the farrier should have a keen eye for measurements and proportions. Always be observant of undue wear and malformation. Be alert for the unpredictable.

This same farrier should learn to work quickly yet at the same time produce a high standard of work, never giving half a dozen blows where one would suffice.

As already stated, a basic knowledge of anatomy is essential as is the quality of shoeing patience.

Awkward farriers make awkward horses and these, especially when they are big, are not easy to work with.

"There are no problems unique to shoeing draft horses. Yet a number of farriers often seem somewhat reluctant to shoe these gentle giants of the horse world..."